WRITE RIGHT

A Practical Guide To Business And Technical Writing

Susan R. Quinn

Illustrations by Regina Hogan

Joanne—
Thanks for being
my friend
Susan

KENDALL/HUNT
PUBLISHING COMPANY
Dubuque, Iowa

Library of Congress Catalog Card Number: 84–81256

ISBN 0–8403–3404–4

Printed in the United States of America

B 403404 01

Contents

Preface, **iv**

Chapter One	**What is a "Professional" Writing Style?, 1**
Chapter Two	**Dealing with Writer's Block (or How to Overcome Terror of the Blank Page), 5**
Chapter Three	**Clear, Concise Writing, 11**
Chapter Four	**Letters, 21**
Chapter Five	**The Planning Process, 35**
Chapter Six	**Editing Your Writing, 47**
Chapter Seven	**Memorandums, 51**
Chapter Eight	**Reports, 59**
Chapter Nine	**Meeting Minutes, 67**
Chapter Ten	**Machine Dictation, 71**
Chapter Eleven	**Punctuation, 75**
Chapter Twelve	**Setting Goals, 87**

Reading List, **89**

Index, **91**

Preface

Frequently during writing courses, participants have asked me for "one good book" on writing. I've always been happy to recommend a number of good books, which are referenced, in part, in the back of this book.

In order to prepare for writing programs over the years, I have also learned and developed strategies for producing effective business writing. I felt that people wanted more than a comprehensive text on writing; they wanted a "nuts and bolts" source which they could refer to, a refresher on the basics of good business writing.

A book that was easy reading, straightforward and sincere was in order. Therefore, I have written this book not only for participants in my courses, but for the business writer who wants practical information which can be used as a guideline, to save time on writing and to produce a better written product.

Since I believe that business writing can be creative, fun and challenging, I followed my own advice and did some "brainstorming" about what I wanted to accomplish in this book. Because my brainstorming was written down not only with the deepest sincerity, but also expressed the essence of what I wanted to share with you, I am including it with this introduction:

I am writing this book because I want people to have a source to go to when they are feeling 'uptight' about a writing project . . . when they want a fresh approach to the everyday demands of writing . . . when they need a refresher on writing letters, memos and reports . . . when they want to make writing a labor of love, not just plain labor.

It's a book for people who enjoy business writing, for those who don't care for it too much, and for those who suffer from 'terror of the blank page.' It's a book which won't make writing more difficult by imposing limitations on your writing, but instead will suggest the flexibility and the formats which can be adapted to any writer's style.

Most of this book is written according to what I perceive as good writing. (No matter what anyone has told you, there are as many "good" styles as there are writers.) My description of good style is limited by my perception of good writing, but I believe that I am flexible enough to encourage you to develop and improve your own writing into a style that is comfortable for you.

I want to express very special thanks to the following people for the many hours they spent editing my book: Judy Riggenbach, G. L. Smith, Martha Warner, and especially to my husband, Jerry, whose editing and support were invaluable.

Susan R. Quinn

1

What is a "Professional" Writing Style?

DEFINING "PROFESSIONAL"

NOTES

First, let us define "professional." This is a definition of a professional: according to *Webster's New World Dictionary,* a professional is one who is "earning one's living from an activity." Therefore, professionals include not only the traditional professions, such as doctors and lawyers; they are also secretaries, supervisors, managers, field inspectors and all the other people who write as part of their jobs.

What is a "professional writing *style*"? Professional style does not fit easily into one definition; however, the term does not suggest that we should all be writing in exactly the same style. Some people feel that professional style implies that our writing should be stuffy and formal. On the contrary! Most writing, except perhaps for formal reports (that essentially only discuss cold, hard data), should be warm, natural and sincere; in the chapters on letters and memorandums, we'll spend more time on the "tone" of your writing.

KNOW YOUR READER

Frequently people think of professional style as one which motivates a response from our readers, and earns us respect. In other words, we think of style in terms of what it does for *us*—not what it does for our *readers*. We should ask ourselves when we write, will the reader understand us? Are our memos read immediately or only eventually, because the reader has difficulty wading through our directives? Do our readers trust our information, because we write *for* them, not *at* them?

NOTES

Deciding what the reader needs to know, and how the reader needs to receive information, is the first step to a professional writing style. We have to decide not only what we are going to *write,* but what they need to *know,* and how we need to explain our information. Some people believe that it's up to

the reader to figure out what we've said, a sentiment obviously not in the best interests of the reader.

NOTES

Is it always easy to define who your reader is? No, it's almost never easy. Sometimes you have a reader who is in your profession and knows your jargon; more often, the reader is much more difficult to define: at times the reader knows the background of a situation, at other times not; sometimes the reader has biases, which you may or may not know about; occasionally the reader is familiar with your profession, but does not have background on the situation; most difficult of all, you may have several readers with diverse biases and amounts of information, who will read the same written product.

Does this mean that writing for the reader is an insurmountable task? Definitely not! In some situations, you may need only a cursory acquaintance with your reader; you, as the writer, must decide just how important the writing project is. An important project, one that has long term ramifications or involves much money, may require that you research your reader, or readers, in depth. In all cases, you must keep the reader in mind.

THE TRADITIONAL PROFESSIONAL

What about those people who are professionals, in the strict, stereotyped definition of the term? These doctors, lawyers, engineers and other professionals are also realizing that the public is demanding that writing, in many situations, be geared to the lay audience. Within the professions of medicine, engineering and law, the competition has become fierce; the professions which deal with the public through writing realize that people will "shop" for the professional who expresses himself clearly, and who works hard to produce written information that can be clearly understood. The public wants to know what is happening, *how* and *why* it is happening; professionals have an obligation, therefore, to be sure that their readers understand them.

2

Dealing with Writer's Block (or How to Overcome Terror of the Blank Page)

NOTES

Have you ever started to write, and found that your mind went blank? Everyone suffers, at one time or another, from terror of the blank page, also known as "writer's block." The reasons for writer's block can be lack of direction regarding your subject, dealing with a difficult situation, or just plain discomfort about writing; you may also have difficulty concentrating, or be afraid of not being perfect and of not having enough information to complete the writing project. Whether you write letters, memos or reports, if you've ever had trouble "getting started," here are some suggestions for overcoming mental roadblocks in writing.

Arm yourself with the proper tools. Everyone should have his/her own dictionary, thesaurus and grammar book. It isn't enough to have those books somewhere in the office. You should have your own books somewhere in or on your desk; most of us will not take the time or make the effort to get up from our desks to pick up a dictionary which is on a shelf across the room. A dictionary, thesaurus and grammar book are current for years; although the rules of grammar and writing are constantly changing, they change very slowly, so these books will be current for many letters to come.

These books are also valuable because they will give you the security of knowing that you are properly prepared to take on any writing project. If a word escapes you, or a question of grammar arises, you are prepared to get an answer quickly. So be sure that you have a dictionary, thesaurus and grammar book close at hand.

NOTES

Make time for your writing. There are several people I know who have trouble writing because they are constantly interrupted by people walking into their offices or by telephone calls. Interruptions to your writing can be distracting, if not downright discouraging. Put aside a time of day when you concentrate the best and when you have the fewest interruptions; for many people, early morning is the best time to write. Some people, however, work on a different body clock; if you are "raring to go" at 6:00 P.M., put time aside then for your writing projects. You know when you are alert and when you are likely to be undisturbed; that's when you should plan to write.

Don't procrastinate. The worst enemy for all of us is procrastination. I know; you may be one of those people who claims that you write best under intense pressure. (If you make that claim, I suspect that you may suffer from masochistic tendencies.) How can you possibly enjoy writing, if you are constantly writing against a deadline, under tremendous pressures? Give your writing projects the same consideration that you give other work projects. Give your-

NOTES

self ample time to create—remember, business writing can be creative—so that you can enjoy the process of writing. Without the pressure of meeting a deadline, you'll find that you suffer less from writer's block.

Forget about writing the perfect letter on the first try. In order to save time, we sometimes tell ourselves that we're going to "get this letter right the first time," so that we don't have to make any changes. That attitude generally causes us to freeze up; we're so busy worrying about writing the perfect letter, that we're afraid to begin writing. Tell yourself that you are just going to think about getting the information down. You'll write a better document if you let the words flow, rather than worrying about every word.

Brainstorm. The "planning process," discussed in CHAPTER FIVE, describes brainstorming in more detail. For now, let's just say that it's important to try to write down all of the ideas that you may want to cover in your writing project, without concern for the order of your ideas, word choice or sentence structure. Brainstorming will remove the pressure of being sure that you've included every single idea in your first draft, and it will help you get started on your project.

When possible, picture your reader. If you know the persons to whom you are writing, get a mental picture of them. If you don't know them, try to conjure a mental picture. It's much easier to write when you can picture your readers and consider their attitudes about your writing topic. You'll think of what they need to know, instead of what you are going to say and how you'll say it.

Be natural. Especially when you are writing letters and memorandums, try to be yourself. Don't twist yourself into knots trying to impress your reader with fancy words. If your readers accept you as an individual, and as a professional, they will appreciate a sincere, warm tone—that sounds like you.

Be objective. In most cases, no matter how emotionally involved you are, it's vital to try to look at all sides of a situation. (You'll notice that I said *try*.) When you are writing to a client who is being difficult, or a client who is disappointed with you, it's important for you to try to stand back from the situation and be objective. If you are enraged about something, don't write the letter. (Or you can write the letter, save it, and probably tear it up in a couple of days.)

There are those who believe that you should write letters while you are still angry, so that you will let your reader know just how you feel. However, letters with inflammatory lan-

NOTES

guage generally make your reader just as angry as you are, which generally results in a non-productive or stalemate situation. If you must write a letter with negative information, you will get the best results by letting the facts speak for themselves. And you will emerge as a person who deals with a problem as a professional. (For more information on writing difficult letters, see CHAPTER FIVE.)

NOTES

Don't be afraid of the review process. If you edit your own writing, try to review your writing as an objective, outside party. Better yet, for your more difficult writing projects, let someone else review your writing. Sometimes letting other people review our writing hurts, since none of us likes to see our writing "bleed." If you are sensitive to the criticism of others, there are ways to make their comments easier to accept.

a. Ask reviewers to use a pencil or black pen, rather than a red pen or pencil . . . seriously. There is no reason why a regular pen or pencil can't be used. Just tell them to make their changes and to put an "X" or check mark in the margin, so that you'll be able to easily locate the line where they've made their notations.

b. Remember that you are not obligated to accept their changes (unless you have a boss who insists). People have a tendency to change any writing style to their own; if someone has done that, stay with your original word choice.

c. Assume that they are revising in your best interest.

d. Ask yourself if their changes contribute to the clarity of your writing; if the changes really help, incorporate them in your final draft.

Relax! If you "know your stuff," and you consider the above suggestions, just take a few deep breaths, close your eyes for a moment—and write away!

3

Clear, Concise Writing

THE SENTENCE—WHAT MAKES A GOOD ONE?

NOTES

There is no such thing as "one good sentence." Sentences, independent of a theme or context, are difficult to evaluate, since a sentence can be clear in one context and vague in another; or it can be too complicated in one paragraph and very poetic in another.

In this chapter we will discuss the elements which, in a well-written context, contribute to good sentence structure. Keep in mind that sometimes we will be discussing the style of a sentence out of context; for that reason we can't be certain of the sentence's definitive meaning. We will, however, try to establish objective criteria for evaluating the sentences we write.

Sentence variety

What is an appropriate sentence length? Some sources say that if you have a sentence shorter than eight words, or longer than 24 words, you may want to take a closer look at your sentence structure. The key to length in sentences is sentence variety: write some long sentences and intersperse them with shorter sentences. By using variety in your sentence length, your writing will have balance and style. Varied sentence length also makes your writing easier to read.

Sentence variety also includes changing the structure of your sentences. If several sentences in a row are written "subject, verb," for instance, the writing will be plodding and repetitious.

Here is an example of both elements of creative sentence variety—variety in length and variety in structure:

NOTES

> "Small companies already employ more than half the workers in this country. In light of hiring cutbacks by the federal and local governments and because the larger corporations are reducing personnel, small companies will provide four out of five new jobs in the future. For that reason, the success of small business has become increasingly important to our economy. However, to do its part in building a healthier economy, small business must find a more sympathetic environment."

Let's look at this paragraph and each of its sentences. First, look at the sentence length. The opening sentence is only 12 words, but did you notice how easy it was to comprehend the paragraph? That's because of the simple, direct, opening sentence. The second sentence is almost too long, with 32 words; however, by punctuating properly, which gives the reader a chance to "mentally pause," the sentence is relatively easy to read. The writer then follows with a sentence of 15 words, and one of 18 words. The paragraph uses good variety in sentence length.

Now, let's study the sentence structure variety. The first sentence begins with the subject (small companies), followed by an adverb (already) and a verb (employ). The second sentence has several consecutive prepositional phrases; the subject (small companies) doesn't appear until late in the sentence. The third sentence begins with only one prepositional phrase (For that reason), followed immediately by the subject (the success). The final sentence begins with a conjunctive adverb (however), and then an infinitive phrase (to do). It's not important to memorize the labels for these words; I've only used the labels to demonstrate the wide selection of sentence structures that we have to choose from in the English language. That variety is what gives our writing balance, flow, clarity and style.

NOTE: You shouldn't worry about sentence structure when you first begin to write. Review your sentence variety and structure when you're reviewing for overall style—which should be late in the editing process.

CUTTING OUT THE CLUTTER

NOTES

We could spend pages and pages discussing words that clutter our writing. Instead, some of the most frequent violations are in this section.

The problem with cutting the clutter out of our writing is that we all have favorite words and phrases. We use them all the time, over and over, because they are comfortable—and because we are too complacent to invent new phrases to replace the old, trite ones. If you see some of your favorite words or phrases in the following section, at least ask yourself if you're committed to writing in a concise, professional style; if so, compile a vocabulary that makes your letter or report special, not just "typical." I'm not suggesting that you write something outrageous or unique; rather, try to write in a style which reflects you professionally and elevates you above the average writer.

The following sentences and their revisions show clutter problems that often occur in writing. In studying the revisions, remember that the strategy is not to have short sentences; the goal is to write descriptive phrases which contribute to the clarity and conciseness of the sentence. Also keep in mind that the revision is not the only way to revise and improve the original sentence.

1A. We hope that at this point in time you will complete the application.

1B. Please complete the application now.

2A. Enclosed herewith please find two (2) copies for your consideration.

2B. Two copies are enclosed for your review.

3A. In accordance with your request we have prepared this report.

3B. As you requested, we have prepared this report.

4A. It is possible that your company may not require this type or detail of study.

4B. Your company may not require a study of this scope.

NOTES

5A. During the course of the investigation, numerous sources of data were considered.

5B. During the investigation, much data was considered.

6A. This is based upon the fact that no evidence was found during the investigation.

6B. This is because no evidence was found during the investigation.

7A. We are also very confident that the service we provide is a fine instrument to enable you our client to rectify any lack of proper maintenance which is currently being provided in your facilities.

7B. We are confident that we can help you correct maintenance problems in your facilities.

8A. Therefore our program of services is limited in scope to address those areas we believe are the most important.

8B. Therefore, our services address the most important areas.

9A. A few additional pieces of information are required to allow us to complete our economic and technical evaluation of this process.

9B. More information is needed for us to complete our evaluation of this process.

10A. As a result of our conversation, we are much clearer on the way in which you handle your projects and what we can contribute.

10B. As a result of our conversation, we understand how you handle your projects and what we can contribute.

11A. It is this organized, efficient management approach that we feel is responsible for our past success on both large and small projects.

11B. This organized, efficient management approach is responsible for our success on both large and small projects.

NOTES

12A. Table 1 shows the two different alternatives.

12B. Table 1 shows the two alternatives.

13A. From my perspective I can see the need for a person to be assigned to the following areas.

13B. I see the need for a person to be assigned to the following areas.

14A. Please let this letter serve as notification that we authorize the placement of the new equipment.

14B. This letter authorizes the placement of the new equipment.

15A. It has recently become evident that the personnel files for many members of this department are out of date regarding special training possessed.

15B. Recently we discovered that personnel files for this department are out of date regarding special training of personnel.

16A. The purchase of the new equipment is necessary in order to provide comprehensive testing of the experimental program.

16B. The purchase of the new equipment is needed for comprehensive testing of the experimental program.

17A. The purpose of this memo is to inform you of the functions of the system, its members and the responsibilities of each member.

17B. This memo describes the functions of the system and its members' responsibilities.

18A. The attached guide was developed to make getting to know the equipment a little easier for everyone.

18B. Everyone will better understand the system by using the attached guide.

19A. It is to be expected that we often forget to regard the life span of the machine, since its reliability is so great.

19B. We often forget to regard the life span of the machine, because it is so reliable.

20A. We have reached the limit of our ability to effectively produce all the work required.

20B. We are doing the best we can!

NOTES

Here is a list of frequently used phrases which clutter and are overused; whenever possible, try to use the more concise version in the right-hand column:

within the realm of possibility	possible
at the present writing	now, presently
due to the fact that	due to
enclosed herewith	enclosed
pursuant to your request	as you requested
in the vicinity of	near
hold in abeyance	wait
for the purpose of demonstrating	to demonstrate
involve the necessity of	require
affording us the opportunity	allowing us
for the reason that	since, because
in the event that	if
prior to the issuance of	before issuing
cost savings measure	to save costs
question as to whether	question whether
will be utilized	will be used
maintain our competitiveness	help us compete
to apply their talents to	to perform, to do

SHUN "TION" WORDS

Words that end with "tion" are what I call formalized verbs, or SHUN words. For example, some people write, "I would like to voice an objection," when they could simply say, "I object." There is nothing grammatically wrong with SHUN words; the problem is that we overuse them:

NOTES

Due to the ramifications of the dissemination of the objections to the qualifications of the definitions. . . .

The reader has to wonder if that sentence ever came to an end. This example of SHUN words is an exaggeration—to most people—I hope. But you can see that overuse of them can make the reader wonder just what the writer is trying to say.

The key is that if you use a SHUN word, ask yourself if it wouldn't be better substituted with a verb (or made part of the predicate). For example:

Determination of the source of the problem was made.

Sounds a little awkward, doesn't it? The sentence reads much better as, "We determined the source of the problem," or "The source of the problem was determined," or, depending on the context, "We found the problem." (Always keep in mind that you never want to revise to the point that you change your true meaning.)

Another example is:

Organization of the energy group was completed.

Why not just say, "The energy group was organized"?

Sometimes a SHUN word is used at the beginning of a sentence to emphasize the actions of "organizing" or "determining"; this can be effective. The point is, *do not overuse* SHUN words. They clutter our writing, tend to make us sound pompous, and do not contribute to the conciseness and clarity of what we are trying to say.

ACTIVE AND PASSIVE VOICE

NOTES

Understanding the difference between active and passive voice is critical to concise, professional writing.

An ACTIVE sentence shows action, and also tells WHO or WHAT completed the action:

We purchased the computer from a new vendor.

That sentence says what the action was (purchased), and who (we) completed the action.

A PASSIVE sentence explains the action, but does not say who or what completed the action:

The error will be corrected by the end of the month.

This sentence says *what* will occur, and when, but it doesn't say *who* will complete the action. Is that important? It's important if you expect Joe to correct the error, and Joe expects Steve to do it, and Steve expects Sharon to correct it. . . .

There are other differences between passive and active voice. Active sentences in letter and memo writing are generally more direct, sincere and natural. For example,

We will help you in any way that we can.

. . . sounds sincere and warm. In contrast, this passive sentence,

Assistance will be rendered under most circumstances.

. . . won't impress most people with your sincerity.

In report writing, passive structure is used in many professions, because it is more formal. Since the primary purpose of reports is generally to disperse information in a formal format, the passive structure is generally accepted. If you are writing a letter or memo, the active voice is more effective because it is more specific (who is doing the action), more informal and more concise; passive voice is vague, formal and less concise. Depending on the writing project, there is a place for both active and passive sentences in business writing.

NOTES

THE MISUNDERSTOOD PARAGRAPH

The paragraph is almost mystical to some writers; its purpose seems to escape those of us who are never sure where to end one paragraph and begin the next one. People misunderstand the purpose of a paragraph; they don't realize that if their writing is well-organized, the transitions will fall into place. (Organizing is covered in the chapter on The Planning Process, CHAPTER FIVE.)

Besides separating ideas or concepts, a paragraph is also a method for breaking up a page; it makes writing more palatable, easier to read. When you are writing, worry about paragraphs during the editing of your writing, not while your writing is in progress.

A criteria for starting a new paragraph is to review your content for a change of pace, topic or direction. If you've made that change, start a new paragraph! When a person must read a document that is essentially one solid paragraph, you've presented that reader with a mental roadblock. That roadblock does not work in your favor if you are writing about an unpleasant or controversial subject. Don't make your reader struggle through or around a written roadblock. If you're writing a letter that will be typed single spaced, block style, limit the length of your paragraphs to two to three inches deep.

By using paragraphs, you help your reader group ideas; if you've organized your ideas well, the reader will find his way from one point to the next.

4

Letters

POINTERS FOR LETTERS

NOTES

Don't you wish there were a form letter for every letter that you need to write? When the pressure is on and a deadline is imminent, letter writing seems to take a lot of time. However, there are some guidelines for writing letters that can help you improve, and in some cases speed up the writing process, as well as assist you in being creative.

Answer letters promptly

The first question you might ask is, "How soon is promptly?" I recommend that letters from clients be answered within two to three days. I know there are reasons for not answering quickly, so let's address those:

"I'm going on vacation."

If you are out of the office for more than a day or two, have someone check your incoming mail. If a client sends a written request for information or needs a response, the person covering for you should either call the client or write a letter, stating that you will be in contact with the client when you return. The response should include an offer to answer the client's needs, if possible, in your absence.

"I don't have all the information yet."

If a client requests information which you don't have yet, don't ignore the client's letter; either call (and document the call in a telephone memo) or write, explaining the delay, and tell the client when you will be able to forward the information. (This is an important time to put yourself in the shoes of your client/reader; *you* don't like to be kept waiting for information or for a service, especially without an unsolicited explanation!)

NOTES

When clients don't make difficult requests, all the more reason to respond to them promptly. Those clients will feel that you are taking good care of them, that you consider them preferred clients, and that you are efficient—and professional.

Don't write to "gentlemen"

NOTES

You may use "Gentlemen" or "Dear Sir" as your salutation when you are writing to an individual, but you probably feel that you are writing to the person as a representative of the company, rather than writing a personal letter to an individual.

However, you sometimes do not know who should receive your letter, and "Dear Sir" or "Gentlemen" is used in place of a person's name. If your letter is not important (and you're the only one who can decide that), it may not be worth your time to find out who should get your letter. An example would be that you are sending out a form letter for some routine material, with no deadline to meet.

If, however, you are writing to a company, and you are concerned about getting a timely response, obtain someone's name to whom you can direct your request. Make a telephone call to find out who is responsible for responding to your request. That way, if you don't get a response, you'll know whom to call. If you write to "Gentlemen" and don't get a response, you'll probably just have to write the letter again—to a specific person.

Put your purpose up front

A business letter should always state the purpose of the letter up front. Keep in mind that you are not writing a novel with a shocking climax. If you have a reputation for keeping the reader in suspense, your letters will probably go to the bottom of the client's incoming mail stack. If he can't determine immediately whether your letter needs priority attention, he may arbitrarily decide that it doesn't.

Be nice

Some people are cold and overly formal in their letter writing, because they think letters are used only to disperse information. Nonsense! Letters can be extremely important for establishing rapport with the client, and letting him know you care about his business. Therefore, you can and should be courteous, sincere, appreciative and warm. Remember that your letters are not only instruments of communication, but marketing tools as well.

NOTES

Stress active voice over passive voice; active voice will contribute to the warmth and sincerity of your letters. (Active and passive voice were discussed in CHAPTER THREE.)

Make it perfect

You must be as sure as possible that your letters have no errors: spelling, grammatical or typographical. Every time an error appears in a letter, you suffer a dent in your credibility (especially if you spell your reader's name incorrectly). If you are not a good proofreader, or if you type your own letters, let one or two others review your letters; most people are happy to do this if you are willing to return the favor. (Editing techniques are discussed in CHAPTER SIX.)

Don't be pompous

Formality in writing was discussed earlier. It is addressed here again because if you are overly formal in your letters, you are destroying your potential strategy for making your client feel special. (I often say that some people write formally to impress their readers; however, if your readers can effortlessly understand what you are saying, they'll be impressed.) Once again, put yourself in the shoes of the reader. How do you respond to overly formal letters? Aren't you annoyed when people "put on airs" with you? Also, some people may feel that you are talking down to them, which doesn't contribute to a positive client relationship.

Distinguish between opinion and fact

If clients need your help, you should always let them know if you are giving them fact or your opinion, not just for their sake but for yours, too. The professional is responsible for telling clients that, after reviewing the facts, he/she has arrived at an opinion. An opinion may be as valid as a fact, but clients have the right to know what you are giving them.

Be honest

NOTES

This is a difficult subject to discuss, because there are people who believe that what the client doesn't know can't hurt him—which may be just another way of covering up a mistake.

Although you should be honest primarily for ethical reasons, you need to be honest for practical reasons, too. (I tell my course participants that clients are like mothers; they may not find out tomorrow what you did wrong, or even find out next month, but they always find out. And you can bet that a big mistake followed by a cover-up is the one that will come back to haunt you.)

We'll discuss how you can tactfully explain a mistake, to lessen the impact of your confession, in CHAPTER FIVE.

Say "thank you"

You probably don't hesitate to tell clients verbally that you appreciate their business. A "thank you" in a letter is even more effective. Frequently, I include a "thank you" at the beginning and the end of the letter—two of the most strategically valuable places to put information. You can say that you "appreciate their business," or that "their help on the project was invaluable," or that "their cooperation helped the job run smoothly." Say thank you in your letters—it is always appreciated by the client.

Typewritten vs. handwritten

In almost every case, I strongly recommend that a letter be typewritten. First of all, a typewritten letter looks more professional. It shows that you took the time to have the letter typed and proofread for the client's comfortable reading. You should never send a handwritten letter if you have terrible handwriting; poor handwriting makes a bad impression just as wearing a wrinkled suit at an interview does.

There is an exception for handwritten letters; I generally *don't* recommend it, however. Some instructors of time management suggest that if you receive a letter that requires a brief response, write your comments on the letter, photocopy it and send it back to the original sender. While that may save time, most professionals with whom I work agree that they would not follow this procedure with clients. Some companies approve of this process for informal, internal memorandums, but that is an internal, corporate decision. The bottom line is that if you are considering sending a handwritten note, try to determine if it will enhance your professional image with your client.

NOTES

Be personal

Don't be reluctant to use personal pronouns. Personal pronouns can be used anywhere in a sentence or letter. If you don't want them to be obvious, you can easily use personal pronouns for a personal touch by hiding them in the letter; don't start more than two sentences in a row with a personal pronoun. Instead of using "I" or "we" at the beginning of the sentence, consider inserting it elsewhere. For instance:

As you know, we have an extensive quality control program.

No one will even notice the "we" in that sentence.

Be careful about using "I." Generally, I suggest that the collective "we" be used; "I" should be used only if you, specifically, are going to do something. For instance:

I will be out of town for the next two weeks.

You simply could not substitute "we" in place of "I."

Generally when you write to a client, you are writing on behalf of the company anyway, so that the collective "we" is more appropriate. Also, "I" should be used with caution for liability reasons. If you submit incorrect information, you could be personally liable, rather than the company being liable.

The following letters illustrate the strategies *not* to use when you write letters. After each letter is an analysis of the letter's problems. Often when you read your letters or those of others, you may sense that something is wrong, but you cannot pinpoint the precise problems. An analytical approach to letter writing will help you identify the organization and word choice which contribute to effective letters. Therefore, we will examine specific problems in the following letters.

NOTES

sent by an independent consulting firm

August 20, 1984

Mr. Robert Blacksmith
Housing Authority and Urban Renewal
1622 Woodmoor Street
Somewhere, Colorado

Dear Mr. Blacksmith:

I am puzzled.

Some time ago we sent you the enclosed letter together with one or more questionnaires advising you of a way to generate cash for your Authority by syndicating your Section Eight Projects and yet we have not heard from you.

The lack of your completed questionnaires has prevented us from submitting to you our proposal. We have by now been able to make proposals to a number of Authorities and have already received a number of acceptances. The receipt of our proposal, of course, does not place you under obligation to accept it. We are, however, reasonably sure that you will find it so unique, complete and also rewarding that you will find acceptance desirable.

My puzzlement is also in trying to figure out the reasons for your lack of response.

How about these reasons and answers: the authority has more money than it could possibly use; all our years in public housing where we spent every day scrounging for more money makes us doubt it. Another reason: by now you have received other proposals—great, but you owe it to yourself to see what we offer; it probably is a better deal for you. Or how is this one: things are so bad that you do not have adequate funds to collect the requested data; here we can be of help. Enclosed please find our check which will partially fund your effort.

If you have any questions please call us, collect, at your convenience. Or if you are attending the upcoming conference in Dallas next month, please call us at the Bloomingfield Hotel where we will be.

Very Truly Yours,

Francis Bracken

Pretty nervy, isn't it? Perhaps the problems are obvious, but let's look at some of them:

NOTES

1. Awkward wording. In the second paragraph, last part of second sentence should read, ". . . has prevented us from submitting our proposal to you." In the same paragraph, last sentence, ". . . that you will choose to accept the proposal."
2. Insulting, inflammatory wording. The next to last paragraph is extremely poor. Mr. Bracken is obviously frustrated because Mr. Blacksmith has not responded to a request for completion of a survey questionnaire. For some reason, Mr. Bracken thinks he can insult his reader into responding. When you ask someone to submit information voluntarily, so that you can submit a proposal for services, insults or intimidating tactics are counterproductive.

Here's a revision of the letter:

August 20, 1984

Mr. Robert Blacksmith
Housing Authority & Urban Renewal
1622 Woodmoor Drive
Somewhere, Colorado

Dear Mr. Blacksmith:

In October we sent you questionnaires related to strategies for generating funds for the Authority; the strategies were related to your Section Eight projects. To date, we have not received a response from you.

Your completed questionnaire will assist us in preparing a proposal for you; the proposal puts you under no obligation to accept our services. We believe that our experience in working with Housing Authorities will contribute to a practical, feasible proposal that will meet your special needs.

After ten years' experience in working with public housing, we understand the many demands on your time. However, if you or a department representative could complete the enclosed questionnaire, and return it to us by January 15, 1985, we will be able to submit a proposal that will have a unique approach to generating income for the Authority.

You are welcome to call me, collect, if you anticipate any problems with meeting the January 15 date. Or if you are attending the conference in Dallas next month, I would be pleased to meet with you at the Bloomingfield Hotel.

Very Truly Yours,

Francis Bracken

This is certainly a more persuasive, even-handed approach to the situation. People are more likely to respond if you emphasize to them the benefits of that response.

sent by an engineering firm

June 15, 1984

Mr. Steve Evanston
BBB Technicians
1434 92nd Street
Everywhere, Colorado

Dear Steve:

Confirming our telephone conversation this date—subject specification is attached for your review and return advise.

Kindly do confirm of your intent to submit to us a proposal to perform these services on a cost-plus-expenses basis.

We would like to meet with your representatives here at our West Coast office on Thursday the 14th day of July at 9:00 A.M.

Our project manager, Mr. Bill Bradley who will be forwarding to your attention, within a few days, technical information such as drawings and sketches for your review. Mr. Bradley can be reached per address and phone number shown below. Additionally-a formal request for proposal with a proposed form of contract will be transmitted to you prior to, or be given to you, at our meeting.

We are interested in your positive response and look forward to meeting with you in this regard on the 14th.

Should you have any questions—or need more particulars—please let either Bill Bradley or me know of such.

Sincerely yours,

Joseph Handel

The only way to critique this letter is to examine it, sentence by sentence. Each problem area of the letter is numbered below; a description of each numbered item follows the text; then a revised version of the letter follows:

June 15, 1984

Mr. Steve Evanston
BBB Technicians
1434 92nd Street
Everywhere, Colorado

Dear Steve:

(1) Confirming our telephone conversation this date—subject specification is attached for your review and (2) return advise.

(3) Kindly do confirm of your intent to submit to us a proposal to perform these services on a cost-plus-expenses basis.

We would like to meet with your representatives (4) at our West Coast office on Thursday, the 14th of July at 9:00 A.M.

Our project manager, Mr. Bill Bradley (5) who will be forwarding to your attention, (6) within a few days, technical information such as drawings and sketches for your review. Mr. Bradley can be reached (7) per address and (8) phone number shown below. Additionally (9)—a formal request for proposal with a proposed form of contract will be transmitted to you prior to, or be given to you, (10) at our meeting.

We are interested in your (11) positive response and look forward to meeting with you in this regard on the 14th.

Should you have any questions—or need more particulars—please let either Bill Bradley or me know (12) of such.

Sincerely yours,

Joseph Handel

(1) Mr. Handel sounds like the kind of man who operates like a machine gun—he works in little spurts, and it's up to associates to figure out how to keep up with him or what he is trying to do. Writing in disjointed fragments forces the reader to "fill in the

blanks." Complete sentences, especially the opening sentence of a letter, are a "must."

(2) What does ". . . return advise" mean?

(3) Kindly? The entire sentence has a pompous ring to it.

(4) It's not clear whether Joseph Handel is proposing that they meet, or confirming a previously scheduled meeting.

(5) The sentence makes no sense with "who" in it. Mr. Handel probably had a thought about explaining who Mr. Bradley was, but by the time he finished the sentence, he didn't complete his original thought.

(6) The commas around ". . . within a few days . . ." are not needed.

(7) "Per" is often used in business writing (per your request, as per your request). The only problem is that it is grammatically incorrect. Per should only be used in measurements, such as nine miles per hour, ten dollars per pound.

(8) "Phone" is colloquial; telephone is more appropriate.

(9) To begin, dashes are made with a double hyphen (—). The dashes in this letter are extremely disruptive. Dashes are punctuation marks which are supposed to grab the reader's attention; they emphasize the information which they surround. Mr. Handel has used them in place of periods, leading the reader to believe that the information which the dashes offset is more important than the rest of the letter. There is no evidence to support that idea.

(10) Confusing and incorrect punctuation—the end of the sentence should read, ". . . transmitted to you prior to, or be given to you at, our meeting."

(11) What is a "positive response"? And what does "in this regard" refer to?

(12) "Of such" is an awkward end to a sentence, as well as to a letter.

Here is a revision of the letter:

June 15, 1984

Mr. Steven Evanston
BBB Technicians
1434 92nd Street
Everywhere, Colorado

Dear Steve:

We were pleased to discuss the upcoming proposal with you. The specification which we discussed is enclosed; we would like to have it returned to us.

Please confirm your intent to submit a proposal to perform these services on a cost-plus-expenses basis.

We look forward to meeting your representatives, as agreed, at our West Coast office on Thursday, July 14, 1984 at 9:00 A.M.

Our project manager, Mr. Bill Bradley, will send you technical information such as drawings and sketches within a few days. Mr. Bradley can be reached at the address and telephone number shown below. Also, a formal request for proposal with a proposed form of contract will be given to you before or during our meeting.

We hope that you decide to respond to the request for proposal and look forward to discussing the work on the 14th. If you have questions before then, you may call Bill Bradley or me.

Sincerely yours,

Joseph Handel

These are the strategies to keep in mind when you write letters. The next chapter describes the process for planning your writing projects and the strategies for writing them.

5

The Planning Process

PLANNING AHEAD

NOTES

Writing is seldom an easy part of anyone's job. If you write letters, you may be one of a large majority who puts off writing projects until the last minute, and agonizes over the final draft, asking, "Will the client understand me? Did I leave anything out? Did I say what I wanted to say?"

If you are questioning the effectiveness of your business correspondence, you are probably allotting too little of your time to planning your writing projects. In reality, you could be using your correspondence as an effective sales and marketing tool. You have to give your business writing the amount of time and planning which it deserves, so that you can integrate your writing with your business strategies. You need to allow sufficient time to organize your letters to meet two important objectives: (1) You must be as certain as possible that your reader will understand your writing, and (2) You should use your writing to project a professional image, that of a knowledgeable, capable businessperson.

If you have difficulty meeting these objectives for business writing, consider using a structured "planning process." The following strategies will show you how to prepare to write your letters. These steps are meant to save time, because you will be so well-organized that you will spend much less time re-writing and editing your letters. The "planning process" includes these steps: (1) determining specifically what you want to write about, (2) "brainstorming" the details of the letter, and (3) using an organizing format.

Theme: limiting the purpose of your letter

NOTES

Write down the theme of the letter, or thesis sentence, to help pinpoint the purpose of the letter. (This is not necessarily the first sentence of the letter.) Sometimes you *think* you know what you are writing about when, in fact, you may not be clear about your purpose. For instance, if you are introducing a new product to a client, you may limit the thesis sentence to, "I am introducing a new product to the client." This is probably only half of your thesis sentence. The other half may be, ". . . and showing the client how he/she will benefit from that product." Writing down both parts of the theme will define the strategy of the letter; referring back to the theme will keep you on track.

It is also very important to try to limit the purpose of a letter to one or two major themes. Too many themes in one letter will dilute the letter's impact and possibly confuse your reader.

NOTE: As you get farther into the "planning process," you may realize that you are digressing from your theme. If you are, stop immediately and decide whether the thesis sentence needs to be revised or if you need to refocus the content.

Brainstorming: collecting ideas

If you labor for hours, even days, over the letters you write, you are probably trying to write the "perfect letter"—the letter which has all of the right points, in the right order, without any mistakes on the first draft. However, a letter rarely is perfect in one draft. For that reason, convert some of those laborious hours into a few minutes of "brainstorming."

Brainstorming is the most complete, stress-free way to organize a letter. First, be broadminded; write down all of the ideas that could apply to the theme. (Don't even bother writing complete sentences.) Second, do not worry about the order in which you write down your ideas. (You can re-organize them later.) Third, wait to write the actual letter until you finish brainstorming—you may be leaving out an important point if you short-cut the process.

This brainstorming technique removes the pressure of making your initial writing perfect. If you experience writer's block, you will find this procedure very helpful; you can begin brainstorming with any idea. Also, you do not have to wonder, in the first stages of the letter, if you have written down every idea or included too much information.

NOTES

Outlining: choose your own format

NOTES

When you see the word "outline," you might assume that you must use a formal outline. Most of us are acquainted with formal outlines, but you can use any method you wish, from flow charts to Arabic numerals. Just be consistent; don't change formats in the middle of the outline.

Once you choose a format, look at the ideas on your brainstorming sheet. Try to select the ideas which will explain the thesis sentence, but don't spend too much time on your decisions at this stage.

Finally, re-organize the "brainstormed" ideas in an orderly, outline format. Put related items together; leave plenty of space to add or remove or relocate ideas. This is the time to be sure that the information is consistent with the thesis sentence. When you have completed the outline, review it carefully. Try to determine if you have left out relevant information. Eliminate ideas that are redundant or irrelevant. When you are satisfied that your information is complete, you are then ready to write your letter.

If you are not accustomed to following a "planning process" for letter writing, this procedure may, at first, appear difficult or time-consuming; remember that changing any behavior or habit can be both.

However, to improve your writing habits and become a more skilled writer, you must be willing to make some changes. Writing is no different from any other skill: to be effective, you must have a positive attitude, be willing to grow and change, and practice, practice, practice.

If the "planning process" is new for you, try to use it with every writing project, no matter how small. (In fact, the shorter the project, the easier the process.) Try to look at your writing objectively. Make the commitment to use your writing as part of your overall business strategy. Well-organized, professional writing will reflect the capable, concerned image which is part of a successful business.

ORGANIZATION AND WORD CHOICE/DIFFICULT LETTERS

NOTES

Once you learn to write difficult or problem letters, you can probably write any type of letter. A difficult letter gives a specific deadline for returning a signed contract, explains how you made a mistake, demands payment of an overdue account, or else! If you learn how to organize these letters, and use objective word choice, you'll be writing letters that gain respect and motivate action.

The keys, then, to effective letter writing are *clear organization* and *objective word choice.* The following will explain these two ideas, and will also include an example of a poorly written letter and its well-written revision.

When writing a letter, one which may put you or the client in an awkward situation, you may tend to back away from dealing with the situation in a clear and forthright manner. You may be tempted to fill the letter with trivialities, fluff and irrelevances, thinking that if the problem is vague, the client will not be disappointed or angry.

Instead, remember that you probably cannot prevent a client from reacting negatively to a bad situation; you *can,* however, lessen the intensity of a client's reaction by being direct and objective in your description of the situation. In that light, here is a method for organizing problem or difficult letters:

(The practical application of these points will be shown in an example that follows.)

1. **Refer to the topic of your letter in the first paragraph or no later than the beginning of the second paragraph.** If you have something unpleasant to write about, you may spend a lot of time discussing information that is not related to the problem. Identify the problem immediately and objectively. Don't go into too much detail, but at least let your reader know what the topic is.
2. **Define or describe the situation.** Sometimes you don't need to give a detailed description or background for

NOTES

the situation; you may have already given the details over the telephone or in previous correspondence. However, when background information needs to be documented, decide how and where the background information should be documented. Otherwise, at some later date, you may need to give an explanation, and without documentation, your reasons may well be considered "unsupported" or "new information." When describing the situation, again be as objective and tactful as possible, whether the problem was caused by you or someone else. "Fingerpointing" (blaming someone else) is generally counterproductive.

When you list items that contributed to the problem, tell the reader which are major (important) and which are minor points.

Finally, eliminate extraneous detail (also known as making excuses); your reader will only be interested in the relevant reasons for the situation.

3. **State your solution to the problem or the benefits of the suggested resolution.** Instead of saving the problem for the end of the letter, place the solution there. (This is usually the part of the letter where people just begin to explain what the problem is.) Although the client may not be pleased with your solution, you may be able to convince him that the benefits, eventually, will be to his advantage.

   This is also the place where deadlines should be stated. If you want people to respond to your letters, give them a specific time frame—"as soon as possible" is not a deadline. Most people are willing to accept deadlines (if they're reasonable deadlines), because they can budget their time accordingly. They will most often meet deadlines if you can tell them what will be gained by their responding on schedule.

4. **Show appreciation.** Let the reader know that you are cooperative, understanding, or empathetic, whenever possible; if your reader is responsible for a difficult

situation, and you see no reason to sympathize, you can always be firm or resolute!

NOTES

Let's look at an example of a problem letter. There is a "before and after" version, so that we can describe the strategy above, and also discuss the importance of "word choice."

July 29, 1983

Mr. Bill Disappointed
YOR Company
12345 Hard Times Way
Nowhere, Colorado

Dear Mr. Disappointed:

Over the years it has been our pleasure to serve YOR Company with the best possible health care. We have always tried to provide the most up-to-date facilities and state-of-the-art equipment. As you know, we have a statewide reputation for service and expertise.

A Mr. John Injured was in our clinic four days ago. (I was just told about the situation yesterday.) He apparently had injured his hand on the job and had come to our facilities for medical attention. Ordinarily we always have a physician on hand over the lunch hour; however, when Mr. Injured arrived, no one was available to help him. Someone should have called me at that time, but no one did. I can assure you that if I had known about the situation, it would not have taken place.

As you probably know by now, Mr. Injured waited two and one-half hours for medical attention. He says that he was in a great deal of pain at the time. We finally came to his aid at 2 P.M.

We are sorry that this situation occurred, but sometimes mistakes happen. Our chief of physicians has assured me that this will not happen again.

Sincerely yours,

Steve Surly

NOTES

That's pretty poorly written, isn't it? Let's look at the letter and evaluate the poor strategy first, and then evaluate the word choice.

Refer to topic—The real problem doesn't appear until the third paragraph, that Mr. Injured waited two and one-half hours for medical attention. The paragraph which should have suggested there was a problem was filled only with fluff.

Define situation—We see what happened, but the explanation includes "fingerpointing" and excuses. Remember, your client only wants a relevant explanation.

State solution—There isn't one. The client should have been told, as a minimum, how the situation would be prevented in the future.

Show appreciation/empathy—"Mistakes happen" certainly doesn't show much empathy. Also, if you are in a position to respond to the client, YOU should probably assure him that you will at least try to see that this situation does not reoccur.

Use appropriate word choice—Even if you just found out what happened, you probably don't want to tell that to the client; that comment only inflames an irritating circumstance. The comment, "He says that he was in a great deal of pain at the time," suggests the writer doubts Mr. Injured's sincerity. Even if the sincerity is in question, that is irrelevant to the situation; it sounds like the writer is trying to make light of the seriousness of the problem.

Here's a revised version of the letter:

July 29, 1983

Mr. Bill Disappointed
YOR Company
12345 Hard Times Way
Nowhere, Colorado

Dear Mr. Disappointed:

As you know, we recently had one of your employees come here for treatment of an injured hand. Unfortunately, Mr. Injured did not receive prompt treatment. As you requested, a description of the situation follows:

Mr. John Injured was in our clinic on July 26, 1983. He had injured his hand on the job that day, and came to our clinic for medical attention. Apparently a physician was not available to help him, and so Mr. Injured waited two and one-half hours for medical attention while in a great deal of pain.

I apologized to Mr. Injured when the situation came to my attention; also, I have sent him a letter of apology. I deeply regret that this situation took place, and have personally taken measures to see that there is not a reoccurrence.

We appreciate your concern, and hope that this letter serves as assurance that we will strive to serve your continued medical needs efficiently and professionally.

Sincerely yours,

Steve Sensitive

It isn't difficult to relate this version of the letter to the guidelines for problem letters.

OPENING AND CLOSING SENTENCES

Starting and ending a letter can be the most difficult part of the task. Here are some suggestions for getting in and out of letters. Remember, the structure of the example sentences is

NOTES

what is important, not the specific content; replace the specific content with information relevant to your own business.

For opening sentences, try these formats:

1. Thank you for taking the time to explain the advantages of your latest model of the XYZ Digit-Locator.
2. As we discussed, I am sending you your updated billing for your daughter's account.
3. As you requested, we are enclosing a duplicate insurance claim form with this letter.
4. In your letter dated January 14, 1983, you explained the reasons for the delay in delivering our new photocopy machine.
5. We appreciate the time you have volunteered for this important project.

For closing sentences, try these:

1. We appreciate your cooperation and assistance in resolving this situation.
2. If you would like to discuss the potential for this new project, please call me.
3. As always, we are ready to meet with you to explain this complex situation.
4. Please call me if you have any problems responding before this deadline.
5. Thank you for answering our questions; we look forward to serving your account.

These sentences are practical alternatives to the overused phrases like, "Enclosed please find . . ." and "If you have any further questions, please feel free to call."

For some people, my suggested approach to writing letters may seem paradoxical. On the one hand, I state that writing letters can be a creative process; on the other hand, the step-by-step approach may appear analytical, and appear to stifle creativity. You must remember that these strategies are meant to be guidelines, a framework, not a set of rigid rules that must be followed. You need to practice using them, too, in order to integrate them with your regular writing habits.

NOTES

6
Editing Your Writing

NOTES

Remember that terrific letter you wrote last week, the one that sounded so professional, clever and confident, the letter you're reviewing *now,* that spelled your client's name wrong and left the second "e" out of "sincerely"?

Unfortunately, the readers of your letters almost surely will not know why a letter isn't edited to perfection, which brings up the next question: Do your letters need to be perfect? The only person who can answer that is *you;* chances are, you may tolerate one error when you're reading a letter sent to you. After the second or third error, you will probably have second or third thoughts about the letter's writer.

That is why you want to send letters that are as close to perfect as possible. Although content is of prime importance, "aesthetics" can make the difference between selling an idea or losing a client. You should send out letters that you can read with pride; letters that reflect your best image; letters that say, in effect, that you take your readers seriously.

The big question is: "How do you find your mistakes *before* your letters are sent?"

Following are strategies for editing your own writing. Not all points will work for everyone. You have to experiment with them, trying them in combinations over a generous period of time, to find out what will work best for you.

Remember, too, that editing is not simply finding typing mistakes. Good writing involves clear expression, sentence variety, concise wording and good organization.

1. When you begin to write a letter, using lined paper, *write on every other line;* if you are typing it, double

space your lines. In this way you'll have plenty of room to make changes. Otherwise, when you try to squeeze the changes in, you may need a magnifying glass to decipher what you wrote.

NOTES

2. Try to *let some time pass* between each stage of the letter—the writing, the editing, the typing, the proofreading. You can be so close to a project that obvious mistakes are easy to miss.

3. *Allow a co-worker to review* your writing. Someone else may spot an unclear idea that you've missed. People are reluctant to "bother" anyone to review writing. If you review their correspondence and they review yours, everyone will benefit.

4. *Read the material aloud.* This isn't easy, especially if you work in close proximity to others. But this is one of the best ways of finding simple words that you

NOTES

NOTES

have left out. When you review your letters in silence, you tend to mentally fill in missing words. This step bypasses that hazard.

5. *Read* the letter *backwards*. That's right; start at the end of the letter, with the last word, and read each word separately. This is a good way of spotting spelling errors—after you've read for content, of course!

6. Learn how to use *"editing notations."* These special signs can often be found in the appendices of dictionaries. They are handy, because you can make changes, but leave the draft of the original wording intact. If you totally blacken out a passage, you may decide later that the wording was really quite good, after all. No matter what anyone tells you, you can't hold your letter up to the light, look at it from the wrong side and figure out what you originally wrote!

7. When *reviewing* your writing *for content only,* and something doesn't "read right," just mark the place and go back to it after you've finished reading the entire letter. Don't get bogged down on a missing word/phrase; it usually will come to you later, when you're not racking your brain for it.

8. *Check for repetitions* in your writing, especially in longer letters. Underline repeated phrases or ideas. When you finish reading, go back to those points and decide if you need the repetition. You shouldn't necessarily avoid repetition; repeating something *can* be a clever strategy for emphasizing an idea.

9. When you've made several changes to a draft, try to *have the draft retyped.* If you've made significant changes, you may have a difficult time seeing beyond the original draft; a clean copy allows you to start fresh.

Give yourself several opportunities to experiment with these steps; when you take your correspondence writing seriously, you can use editing skills as a strategic tool. Take responsibility! Writing is a serious business and all parties involved, from inception to final product, should take pride in the writing and editing process.

7

Memorandums

NOTES

If you think memorandum writing is different from letter writing, you are only partly right. Many of the ideas discussed in CHAPTERS FOUR and FIVE apply to memo writing as well. One of the greatest differences between letters and memos is your audience/reader.

THE AUDIENCE

In this book, memos refer to internal company correspondence. Often you know an internal audience better than you know an external (client) audience. How does that knowledge help you write memos? Remember that the better you know your audiences, the easier it is to write for them; assumptions can be made about their understanding of corporate procedures and policies; you may know some of your readers not only professionally but personally, too.

If you know your reader, and you are writing an internal memo, you can usually write less formally for that reader. You do not need to establish credibility with the reader by using a formal style, if what you are writing is honest and justified. Some people tell me that they feel comfortable writing in an informal style to their peers, or even to their immediate superiors. However, when they write, for example, a memorandum to the president of the company, they suddenly feel obligated to use an overabundance of multisyllable words and very long sentences (also known as gobbledygook).

Remember, the president has many memos to read besides yours. He/she probaby does not want to wade through an overly formal memo with a cluttered message. In other words, the higher up your memo goes, the more obligated and committed you should be to making memos clear and concise.

WHEN TO WRITE (AND WHEN NOT TO WRITE) A MEMO

NOTES

Are you a person who thinks writing memos is a priority, often above other job responsibilities? Then you may be writing memos when you should be doing something else. Especially in large corporations, most employees receive several memos every week; reading them can take a lot of time.

Therefore, most people discriminate about the memos they read, and when they read them.

NOTES

Guidelines for writing a memo are the following:

Can I handle this more easily, more efficiently, by telephone? If you know your readers are responsible and responsive to your comments/requests, a telephone call may be just as effective.

Does your information warrant a memorandum at all? If your readers don't really need the information, don't clutter up their reading time.

You're angry about a change in corporate policy—should you write? If you were not in a position to give input before the change was initiated, you're better off to "wait and see." Sometimes a change that initially appears to be a poor one can turn out to be a good one.

Don't broadside. If you are writing a memo to reprimand, be sure that you don't include everyone in your comments. For instance, if some people are repeatedly taking long lunches, don't send a memo to all employees, criticizing the lack of promptness—unless, of course, you acknowledge the people who DO observe the rules.

Write with forethought. If you're asked for an opinion about something, take enough time to think it over. There are times when you are asked for information "yesterday." Try to get a reasonable deadline when you need to submit information or opinions that will require careful or substantial research or thought.

Don't write for revenge. Writing in anger is generally counterproductive. Give yourself time to cool off and reflect on whether writing a memo is your best strategy. For example, you probably know of at least one person who was promoted into a position that someone else had earned, or that someone else was better qualified for. The person who should have gotten the promotion might be tempted to write a scathing memo to a supervisor or manager, about the unfairness of being passed over. However, the person who did receive the promotion may decide he/she does not want the position, may prove unqualified, or may even lose the job. If you write an angry memo about not receiving a promotion, you may permanently put yourself out of the running.

NOTES

Can an assistant or secretary write the memo? Many memos may be routine, and don't require your personal input. If a secretary is qualified to write your memos, encourage him/her to do that. Also, have an assistant you are training take on more responsibility; get that person involved with writing projects. (Several people whom I supervised over the years grumbled when I delegated writing projects; later they

thanked me for sharing that experience with them, including my comments on the writing they generated. The experience helped prepare them for moving ahead in the company.)

NOTES

DOCUMENTING TELEPHONE DECISIONS

Why not just pick up the telephone when you need to communicate with a client? Sometimes the telephone *is* the best way to discuss everyday issues with others. But you must decide how important the item of discussion is. Too often, a decision is made or verbal agreement is given without backup documentation. Later that decision may come back to haunt you, because you remember the client's comments differently than he/she does.

You need to determine when to document a discussion. Some people will use the telephone initially, but follow up with either a telephone memo to be filed internally, or a follow-up letter to the client to reiterate the discussion and decisions that were made. Just remember that the need for documentation is the criteria for the decision to follow up in writing.

EXAMPLE MEMORANDUM

The following example demonstrates the problems a reader can encounter when a memo is badly written, particularly when the description of the topic is ambiguous and the style is overly formal:

> "To ensure that each engineer has the maximum opportunity to take an active role not only in his project activities, group, and division objectives, we will place special emphasis in developing your monthly work assignments by periodic meetings between the staff group supervisors and his engineering personnel. The work assignment form which is attached has been revised to provide you with a better opportunity to plan and then evaluate the utilization of your time.
>
> "It is hoped that through your participation in developing your monthly work assignments, we can improve our overall

division performance in this year and provide maximum growth opportunities for all division personnel."

NOTES

First, let's look at the grammatical problems. In the first sentence the writer uses the phrasing, "not only"; that phrasing is not complete, unless it is followed later in the sentence with, "but also." In addition, in the first sentence, there is a lack of agreement—"staff group supervisors" should be followed with "and *their* engineering personnel."

Look at the biggest problem with this memo. What is the purpose of the memo? To introduce a new work assignment form? To help people use their time better? To improve division performance? All of the above? If so, in what order of importance?

The writer didn't take the time to choose his primary purpose. If you're not sure what you're writing about, the reader is going to have a difficult time figuring out your purpose.

Another problem is trying to judge the tone of the memo; does the writer sound concerned? Does he try to relate to the reader, or does the writer sound formal and distant? When you want people to adapt to a change in procedure, simplify the process by being considerate, sympathetic and empathetic to the adjustments they will need to make.

Taking into consideration an *early statement of purpose, sincere direct tone,* and *sentence variety,* here is one way the memo could have been more effectively written:

"We have worked with group supervisors to set the following goals for this year: (1) to provide methodologies for engineers to improve job performance, and (2) to improve overall division performance.

"We are implementing a new work assignment form. The form will help you to better plan and evaluate the time you spend on the job. In addition, engineers will meet with their supervisors at least once each month to develop revised work assignments.

"By using this new form, and by encouraging frequent, open communication between engineers and supervisors, we look forward to improving individual and group productivity and performance in 198_"

NOTES

This version of the memo states immediately that the memo's purpose is to discuss improvement of individual and division performance. The sentence structure is concise and direct; the tone is sincere by using the personal pronoun, "we," instead of the more formal, passive, "It is hoped." For these reasons this memo will meet less resistance than the first version.

In summary, be sure that you know precisely what your primary purpose is when you write a memo; then determine what your secondary and supporting purposes are. Good grammar and punctuation skills will enhance your credibility, which is helpful when you are trying to persuade people to cooperate with a change in procedure. Finally, create a tone and use language which lets the reader know how you feel, and how important your information is.

8

Reports

The book, *Report Writing* by Graves and Hoffman, best describes what a report is:

NOTES

> *Every report is expected to communicate useful information. Information is more than vague ideas, feelings, opinions, and prejudices; it is verifiable fact or at least the conclusions of recognized specialists. The information in a report may come from a variety of sources: direct observation and experience, texts and experiments, surveys involving interviews or questionnaires, unpublished records or reports, and published materials. What makes such information* useful *is the kind of conclusions or recommendations that it leads to. Primarily information in a report is not introduced to amuse, to entertain, or even to educate. It is always intended to have a specific use for its reader: to facilitate work he is doing or to help him make important decisions.*

Nearly everyone writes reports, whether they realize it or not. A memo that is longer than one page may actually be a report. When there is a large amount of content, a formal approach to organizing and presenting that information is essential for an effective product.

GENERAL REPORT FORMAT

Every profession has a style guide for writing reports. For this book, I have selected a generic format for reports. The titles I have chosen for each section are arbitrary; you can use any heading or sub-heading which is appropriate for your

NOTES

reports. (There are also suggested alternate headings, and they are just that—suggestions.)

Most reports will include the following sections, depending on the length/complexity of the report: Table of Contents, Introduction, Summary of Findings and Recommendations, The Body, Detailed Findings and Recommendations. The following description shows the order in which your information should be presented in your report. *This is not the order in which the report can/should be written.* We will discuss the strategy for writing the report later.

Table of Contents

When do you need a table of contents? If the table of contents looks ridiculous for the length of your report, you probably don't need one. Generally, I suggest that if the report is at least seven to ten pages long, or has more than three sections, you may include the table of contents. The table of contents is incorporated to assist the reader in locating information; if the report is short, the reader probably won't need the table of contents. (This requires putting yourself in the shoes of the reader again.)

Introduction (Purpose, Background, History)

Depending on your profession, the introduction can include any of the following:

- Reasons for writing the report
- Why your company was chosen to write the report
- Location for the investigation
- Who was on the study team
- How the study for the report was conducted
- When the report was written

Remember that the reader of the report may be reviewing several reports at the same time. Don't assume that the reader will figure out what you are writing about, or even remember why you are writing the report, if you don't include an introduction.

NOTES

Summary of Findings and Recommendations (Conclusions, Summary, Executive Summary)

This section of the report tells the reader "up front" the most important findings and recommendations, or conclusions of the report. Sometimes your readers will not have time to read the entire report, and they need to know the "bottom line" of the investigation. (An engineer told me once that he would not include a summary in his reports, because he felt that the reader should have to read the entire report, after all his efforts to put out a quality product. He obviously did not have the reader foremost in mind.)

Some companies are asked by clients to write an executive summary which is bound separately from the report. This is a valuable tool for the client who receives a very long report and may have little time for, or interest in, reading the entire report.

Summaries are difficult to write because we need to choose a limited number of points which reflect the most important findings and recommendations. The writer, then, must be very selective about the points included in the summary. Also, be sure that the recommendations relate to your findings.

If the detailed findings and recommendations (which are placed near the end of the report) are limited, you may not need a separate summary section near the beginning of the report; duplicating information in both sections is unnecessary. Determine where to put a limited findings and recommendations section, and when one is needed. Sometimes one or two sentences at the end of the introduction section will be sufficient summary, in place of an actual summary section.

The Body (Methodology, Data Collection, Approach, Analysis, Investigation, Assumptions)

NOTES

The body of a report is a description of the methodology used to reach your findings and recommendations. It is the substance of the report, and validates the conclusions you have reached. The body will often include sub-headings. (Later in this chapter are the strategies for organizing the information in the body of the report.) The body may also describe the limitations of the report, as well as subjects which are covered in depth, to substantiate the investigation or analysis.

Detailed Findings and Recommendations (Conclusions)

This section lists the findings and recommendations which are a result of the investigation described in the body. This information can be put in a single section of the report, or you might have a findings section and a recommendations section. Whether the information is included in one or two sections, the findings and recommendations must address the information covered in the body of the report.

Frequently, for longer reports, there are readers who may be interested in reading only certain sections of a report; therefore, they will look for the findings and recommendations which apply to those sections of the body. In very long reports (at least 75 to 100 pages), findings and recommendations can be put at the end of each major section of the body. All sections of the body in longer reports are then self-contained, including their respective findings and recommendations.

Attachments (Appendices, Exhibits, Figures, Diagrams, Tables, Charts, Graphs, Calculations)

Do these always go at the end of the report? This question is always debated and not easily resolved. A pet peeve of mine occurs when I am reading the narrative of a report; at the end of a page, I am in the middle of a sentence, and I turn the page to complete the sentence—and a table fills the next

NOTES

page—and the next page—and the next page. By the time I find the rest of my sentence, I can't remember what I read in the first place. If this occurs often enough, I become very irritated. And if the writers of the report are trying to persuade me about their approach to an idea, they have already created a negative impression.

When using tables and charts, put yourself in the reader's place: does the reader need this information in the text, or is it supporting information that can go in an appendix? Is a chart or diagram the best way to explain a methodology, or would a narrative be better? Sometimes it's worthwhile to simply ask the readers whether they want tables and charts attached or if they want them in the body of the report. If they have strong feelings one way or the other, you're better off knowing before you organize the report, not after the report is submitted.

One suggestion for those who feel a table is required in the body of the report is to compose a "summary table" which reflects only the most important data. Often a summary table fits neatly into the body of narrative, without greatly disrupting the text.

THE STRATEGY FOR WRITING THE REPORT

We've already discussed the importance of brainstorming and outlining. Too often, a great deal of time is spent collecting the data for reports, and very little time is left to organize and write. The following sections describe the strategy for organizing and writing the report.

Some people write the Introduction first. This is one way to clarify the purpose for yourself. However, be prepared to rewrite the Introduction, because as you organize and write the report, the Introduction may need to be changed.

If you don't write your Introduction first, most people organize and write the Body of the report. These are the steps to follow for preparing the Body:

1. Review the collected information. After reviewing the information, brainstorm, on paper, what you perceive are the major subject areas for the report. Leave plenty of room for additions and deletions. The major areas you list will become the major headings for the report.
2. Reorganize the major headings in the logical sequence for the report.
3. Break down the major headings into sub-sections, but don't worry about writing the sub-sections in sequence.
4. Determine a criteria for the sequence of the sub-sections; for example:
    - A history section will be chronological.
    - Items listed first or last will leave the biggest impressions.
    - Methodology will be described step by step.
5. Review the order in which you have listed your major headings and sub-sections—do they still make sense in that order? Rearrange them, if necessary.
6. Make a detailed outline from your brainstorming—don't start writing the report yet!
7. Review your detailed outline. Is it complete? Have you included irrelevant information?
8. Write the body.

NOTES

Sometimes there is a preferred format which a person in your company, or the client, expects. Be sure to adapt the organization of the report accordingly.

The next section which might be organized is the Attachments; very often it is worthwhile to determine, as you are writing the Body, which tables and charts will be needed.

After organizing the Attachments section, write the Detailed Findings and Recommendations. Again, be sure that this section addresses the information covered in the Body.

NOTES

Either list all the findings, then all the recommendations; or list a finding, immediately followed by a recommendation (when applicable). The strategy for organizing this section depends on what you think is the most appropriate way to present your information.

After the Detailed Findings and Recommendations, write the Summary of Findings and Recommendations. Be sure to limit this section to the major findings and recommendations.

Next, decide if you need a Table of Contents. Finally, either write the Introduction (if you didn't write it when you started), or review the Introduction to see if it needs to be rewritten.

When you have finished writing the report, allow sufficient time for reviewing the report for continuity, logic, clarity, completeness, redundancies, style, and for typing mistakes. Is the report well-organized? Should you have a disinterested party review the report before you submit it? Is it visually, aesthetically pleasing, with plenty of white space (wide margins, space between paragraphs), properly placed charts and tables, frequent paragraphing?

Just remember, whether submitting a report for internal consumption or submitting it to a client, every report you write makes a professional statement. That report will sit on the client's shelf or a supervisor's desk and be a reminder of the quality of work you have done for him.

9

Meeting Minutes

NOTES

Nobody likes to take minutes during a meeting. You have to try to listen and write at the same time; you're never certain about how much you should write down. However, meeting minutes can be critical to efficient operation and reduction of liability for a company.

First, establish criteria that your firm can use for taking meeting minutes to insure that everyone will include at least a minimum of information. A set criteria using the following ten guidelines can help professionals establish good listening and writing habits, when dealing with the client at meetings. Also, once a routine is established, you can take more efficient and relevant meeting minutes which are easier for your typist to transcribe back at the office.

1. Always include the date and location of the meeting in your minutes. Many firms have a pre-printed form for the transcriber's convenience. It's important to include date and location because you may be working on more than one project with the client—or your client may have projects in progress with other firms.
2. List the client's representatives, your firm's representatives, and all other attendees who impact the project. Even if you think that an attendee is only a bystander, be sure that all people are identified; you can later determine their status on the project, and whether they should be referenced in the minutes.
3. Clearly identify the project, for your information and the client's. Both of you probably have many other projects. Also, related projects sometimes have similar names, so accurate identification can prevent future confusion.

NOTES

4. Rather than write the meeting minutes in narrative style, number each of the items discussed. It's much easier to read items listed numerically, and more efficient for checking the notes at a later date.

5. Use active, not passive, voice! If the passive voice or indirect form is used, you may not be able to determine who actually made a statement.

   Use the active voice in writing—"Bill Watkins said that we should proceed on the project," not "It was agreed that we should proceed on the project." Never assume that anyone will remember "who said what" at a project meeting.

6. Whenever there is disagreement/agreement on an issue, indicate any exceptions to decisions. That means, list the person by name. Again, if only the firm's name is listed, you're not being specific. If you're unable to get the client to commit himself on an issue, it's important to indicate that no commitment was made.

   Since meeting minutes may be submitted to the client, try to be as objective as possible. The client may be indecisive, uncooperative, even unknowledgeable about the intricacies of your profession—but he's still your client.

7. List items which you or the client must "follow-up" on, or "action items." Responsibility for those items is then clearly defined, and the meeting minutes will be a reminder to you and the client about items which need to be completed.

8. The adage, "More is better," definitely does not apply to writing meeting minutes. Items discussed which affect the project should all be listed. But the details of each discussion do not have to be written down. Just describe items so that readers of the minutes can readily identify them. Indicate disagreement/agreement/decision, and do not editorialize.

9. Write the draft of your notes no later than the next day. That way, you are less likely to forget points that

you might not have written down. If you can't remember something, or some parts of the notes are not clear, ask another attendee as soon as possible.

NOTES

10. Remember that when you initially record the meeting minutes, you are writing only a rough draft. The original does not have to be—in fact, probably will not be—perfect.

The final draft should be typewritten in clear, concise sentences. Send a final copy to the client, and add a note to the minutes stating, for example, that you will assume that the minutes are acceptable as submitted, if you do not hear from the client.

10

Machine Dictation

NOTES

People with whom I work are sometimes surprised that I think dictating machines are wonderful tools; some people feel that talking to a machine compromises the creative process of physically putting pen to paper. However, the physical act of using a pen does not insure a quality product; also, most of us can dictate faster than we can write. You should save time on your writing whenever you can, and a dictating machine is one time-saving tool.

In addition, many people have difficulty using dictating machines; few of us are really comfortable talking to machines, except in our most desperate, lonely moments. However, if you can save time overall on writing projects, and get past an aversion to communicating through machines, the following suggestions should make dictating to a machine much easier.

Know your machine. Find out how all the buttons work; play with the machine, experiment. Best of all, work with someone who knows how to use the machine, so that you're not overcome with frustration before you really have a chance to get acquainted with the machine.

Make an outline. Take the time to brainstorm and organize your ideas. Very few people can clearly and succinctly dictate "off the top of their heads." When you jot down important points, you'll be able to concentrate on sentence structure and word choice.

Make the first draft a rough draft. Don't try to dictate the perfect memo. (We discussed this under "Writer's Block," CHAPTER TWO.) Let the secretary know that you are dictating a draft, which you will review and edit. With this understanding, you won't use up expensive stationery for drafts.

Visualize your reader. Pretend that the person you're dictating for is sitting in your office, and that you are sharing some ideas with him/her. (Although this may sound like a silly procedure, you have imaginary conversations all the time—asking a boss for a raise, giving an acceptance speech, lecturing an employee on a mishap). Just use your imagination.

Get your secretary's help. Don't expect much cooperation from your secretary if you have bad dictating habits. I can't count the number of secretaries who tear out their hair because a person chews gum while dictating; or people let the tape go on and on without saying that they have completed

NOTES

their dictation. Secretaries can often give you hints on how to make machine dictation easier for you and for them.

NOTES

Start on something small. Don't try to dictate a 50 page report on your first effort. Choose short letters and memos for your first several projects. As you become more comfortable and familiar with the process, you can work up to more complicated projects.

Punctuate verbally. Let the secretary know, as often as possible, when to begin new paragraphs and where to place punctuation.

Practice, practice, practice. The only way to become comfortable with a dictating machine is to use it—often.

If you remember that a memorandum or letter should be an efficient tool for communicating, a way to share information with the people with whom you work, your memos and letters, whether created with pen and paper or by machine dictation, can effectively represent you personally and professionally.

11

Punctuation

NOTES

This chapter discusses most of the punctuation which is used in business writing; it is not meant to be a comprehensive dissertation on punctuation, but rather a description of the most common usages and rules of punctuation. The chapter is arranged alphabetically, from "apostrophes" to "slashes."

APOSTROPHE

Use an apostrophe to show possessive case of nouns and indefinite pronouns.

1. For singular nouns and indefinite pronouns, add an apostrophe and "s." (One, everyone, someone are indefinite pronouns.) NOTE: If a singular noun ends in "s," the second "s" is optional.

   Ted *Jones's* house or Ted *Jones'* house

2. Do not use an apostrophe with the following possessive pronouns:

   hers, its, theirs, yours, ours, his, whose.

3. Add only the apostrophe to plural nouns ending in "s." Add the apostrophe and "s" to plural nouns not ending in "s."

   girls' choir   *men's* club

4. Add apostrophe and "s" to the last word only of compound words or word groups.

   my *mother-in-law's* car   *someone else's* department

5. Add apostrophe and "s" to each name to indicate individual ownership.

George's and *Albert's* houses

NOTES

the *architect's* and the *contractor's* offices

6. To show omissions in contracted words or numerals use the apostrophe.

   cannot—*can't* class of *'66*

7. To form plural of lower case letters and of abbreviations followed by periods, use the apostrophe and "s."

   To prevent possible confusion, use an apostrophe and "s" to form plurals of capital letters, of symbols, of abbreviations not followed by periods, and of words referred to as words.

   The word "bookkeeper" has double *o's, k's* and *e's.*

   The positions were offered only to *Ph.D.'s.*

   He earned four *A's* this quarter.

   The machine misprinted the *17's* on each page.

   The weight was indicated by *#'s.*

   Local *AFL's* and *CIO's* concurred.

   There are five *and's* in this paragraph.

The following examples show that the placement of the apostrophe can be important for clarity:

Those puppies look like *my cousins'.*

Those puppies look like *my cousins.*

BRACKETS

Use brackets to set off explanations or editorial comments in quoted matter and, when necessary, to replace parentheses within parentheses.

The letter from Mr. Jacobs said, "For this project we will need your cooperation *[probably Smith's and mine, he means]* to see that the work is finished on schedule."

Perhaps Strunk and White's little book should be required reading for all writers (The Elements of Style *[New York: MacMillan, 1979]*).

NOTES

COLON

The colon means "namely." Therefore, use the colon to precede a restatement, an explanation, a list of particulars, an appositive, an amplification, or an illustrative quotation. A set of examples follows.

A restatement: *He supports our ideas: he wants the plan to succeed.*

An explanation: *The graph defines the areas of growth: at the top we see the peak months and at the bottom the slow starting months.*

A list of particulars: *He collected what he needed for the report: the photographs, the computer data, the letters from experts, and the notes from his own extensive research.*

An appositive: *He had selected the best man for that position: Herb Stone.*

An amplification: *The purpose of the organization was announced: to provide aid to the financially disadvantaged.*

An illustrative quotation: *Mr. Barker requested agreement on this issue: "A house divided against itself shall not stand."*

The colon appears after an independent clause; it is not used immediately after a verb or after the phrase "such as."

COMMA

The rules for commas are actually related to verbal sounds; when a pause or a variation in voice pitch is indicated by the structure of the sentence, the comma is used.

NOTES

1. Commas are used to separate independent clauses. (The comma appears before a coordinating conjunction to separate independent clauses. Coordinating conjunctions are *and, but, or, nor, for, so,* and *yet.*)
2. Commas are used to separate coordinate items in series. The use of the comma before the final item in a series is being dropped.
3. Commas are used for reading clarity.
4. Use the comma to enclose elements which interrupt the sentence pattern.
   a. Use commas in pairs, if in mid-sentence.

      He could revoke his decision, however, if new facts were presented.

   b. Use commas singularly if the element is first or last in the sentence.

      However, he could revoke his decision if new facts were presented.

      He could revoke his decision if new facts were presented, however.
5. Use commas to separate words or groups of words which are grammatically unrelated to the rest of the sentence.
   a. Interjections

      My goodness, what happens now?

   b. Nominals of address

      You see, Henry, we want your ideas.

   c. Dates and addresses

      We plan to meet again, January 15, 1985, at Sable's restaurant, 5000 South Main Street, Omaha, Nebraska.

   d. Introductory phrases and adverbial clauses

      As you might imagine, the board rejected the idea.

      While he waited, he reread the schedule.

NOTES

e. Titles or academic degrees following names

John Harper, D.D.S., spoke about recent innovations in his profession.

f. Transitional words and phrases

Garth knew, therefore, that he must alter the procedure.

g. Attribute expressions in dialogue

"You must go," he said, "before the meeting begins."

h. Absolute constructions

The car, its tires screeching, pulled into the gas station.

i. Contrasted coordinated elements with *not*

It is John, not Henry, who wants to go.

j. Non-restrictive and restrictive elements (clauses and phrases)

Non-restrictive elements are not essential to the meaning of the main clause and may be omitted. They are considered parenthetical, and are therefore set off by commas. Restrictive elements follow and limit the words they modify. They are essential to the meaning of the main clause and therefore are not set off by commas.

(i) Adjective clauses

My mother, whom you all know, will be there. (non-res.)

My aunt who lives in L.A. will visit this summer. (res.)

(ii) Appositives
An appositive is a noun or noun substitute set beside another noun or noun substitute to identify or explain it.

My brother John visited me last week. (restrictive: he has more than one brother and we need to know which one visited him.)

My brother, John, visited me last week. (non-restrictive: he has only one brother and the name is just additional information about him.)

NOTES

DASH

The dash should be used sparingly.

The dash is the opposite of the hyphen.

The hyphen holds two pieces together:

He is twenty-two years old.

The dash pushes two things apart:

His father—not his mother—supported his efforts.

1. The dash sets off parenthetic material the writer wishes to emphasize.

   It was Clark's election to chairman of the board—not Clarence's resignation—which upset Frank today.

2. The dash indicates a sudden break or change of thought.

   We'll need to order more stationery—oh, did you call Howard about tomorrow's lunch?

3. The dash indicates an unfinished or interrupted statement and faltering speech in dialogue.

   What in the world—?

   But, I—you see—I tried, Mr. Lawrence, I tried—but—but, she wouldn't answer the door.

4. The dash sets off an introductory series from the main part of the sentence that explains or amplifies the series.

   Brainstorming, outlining, drafting, editing—these are all essential steps to good writing.

ELLIPSIS

NOTES

An ellipsis mark is three spaced periods used to indicate omissions from quoted passages. If the ellipsis occurs at the end of the sentence, add the fourth dot or period for the end punctuation, and enclose the ellipsis within the quotation marks.

EXCLAMATION POINT

Use the exclamation point after an emphatic interjection or after a phrase, clause or sentence to show surprise, disbelief or other strong emotion.

HYPHEN

1. Use the hyphen to join two words as a single adjective preceding a noun, e.g., *well-organized.*
   Adverbs ending in "ly" do not form compound adjectives:

well-known engineer	but: *widely* known engineer
fast-moving belts	but: *rapidly* moving belts

2. Use the hyphen to divide a word at the end of a line; this division should occur between syllables or between consonants.
3. Use the hyphen to separate the prefix ending in a vowel from the basic word beginning with the same letter, e.g., *re-establish,* or to separate a prefix from a proper name, e.g., *un-American.* Sometimes with usage these hyphens may be dropped, as in "cooperate."
4. Use the hyphen to combine written numbers as in *twenty-one* through *ninety-nine.*
5. Use the hyphen in the expression of time in words, e.g., *eight-thirty.*
6. Use the hyphen to indicate inclusive dates, *1950–1960,* or inclusive pages, *1–10.*

7. Use the hyphen to indicate fractions written in words: *one-half, three-fifths.*

NOTES

Adjectives which should be hyphenated before a noun should not be hyphenated in the predicative position (verb's part of the sentence):

Dr. Sauer is a *well-known* physicist.

Dr. Sauer is *well known* as a physicist.

ITALICS

(If you are using a machine which does not have an italics feature, underline the titles of the items listed below.)

books	works of art
periodicals	musical works
plays	movies
long poems	names of ships, aircraft
pamphlets	foreign words and phrases
newspapers	

1. Italicize words, letters, figures spoken of as such or used as illustrations.

   "Many school children find it hard to remember when to use *they're, their* or *there* correctly."

   "Niece is an example of an *i* before *e* except after *c* rule."

   "The *2* and *4* on her mailbox were reversed."

2. Italicize a word to be emphasized.

   "He *must* accept the justice of the court."

PARENTHESES

1. Use parentheses to enclose inserted material that supplies relatively unimportant data (or mere asides).

NOTES

Parentheses play down the importance of what they enclose; they are most appropriate for things mentioned in passing.

2. The opposite of the dash which sets off material to emphasize it, the parentheses set off material to de-emphasize it.
3. There are four types of parenthetic indicators: square brackets, round brackets (parentheses), dashes and commas.
4. Punctuate after dashes and brackets as if they were not there. When a wholly detached expression or sentence is parenthesized, the final punctuation comes before the last mark of parenthesis.

PERIOD

Use the period after declarative or mildly imperative sentences, after indirect questions, and after most abbreviations.

QUESTION MARK

Use the question mark after direct but not after indirect questions.

QUOTATION MARKS

Use quotation marks for the following:

1. To enclose direct quotations.
2. To indicate names of short works, titles of TV shows.
3. To set off technical words or words used in a special sense.

   John asked for an explanation when his attorney referred to the "exclusivity" of the agreement.

   His "proposal" was hardly an outline.

4. To set off a word or expression that is more formal or less formal than the context of the particular passage.

   His "hokey" demeanor offended the receptionist.

5. To indicate foreign words and phrases.

   The judge referred to the "writ of habeas corpus."

6. To emphasize familiar words used in unfamiliar ways.

   We watched the inexperienced skier "crash and burn" on the advanced slope.

NOTES

Punctuation with Quotation Marks

1. Use a single quotation symbol (') for quotation within a quotation.
2. Commas and periods conventionally precede a final quotation mark.
3. Semicolon and colon conventionally follow a final quotation mark.
4. A dash, question mark and exclamation mark go within the quotation mark when they apply only to the quotation; they are placed outside the quotation marks when they apply to the whole sentence.
5. A terminal mark is not duplicated at the end of (outside of) a quotation even when logic might seem to require its repetition.

SEMICOLON

1. Use the semicolon between main clauses not joined by a coordinating conjunction—*and, or, nor, but, for, so, yet.*

   His proposal was presented clearly and the board will decide tomorrow.

   His proposal was presented clearly; the board will decide tomorrow.

2. Between coordinate elements containing commas, use the semicolon.

NOTES

After careful preparation, John presented his proposal clearly; and, even though John expects its approval, the board will not decide until tomorrow.

3. Do not use the semicolon between parts of unequal grammatical rank.

   Incorrect: *After careful preparation; John presented his proposal clearly.*

   Correct: *After careful preparation, John presented his proposal clearly.*

4. When a listing of three or more elements separated by commas occurs, use the semicolon for clarity.

   Representatives to the committee come from three firms: Brooks, Caplan, Hartsell of Monarch Company; Lawrence, Jensen, Taylor of Rubin, Inc.; Jefferson and Donaldson from PBC.

SLASH

This mark of punctuation is used to indicate an option.

We hope to hire a new bookkeeper/accountant for our office.

12

Setting Goals

NOTES

By now, I hope you agree that your writing is a personal and professional reflection of you. With the best of intentions, it is still extremely difficult to change your writing habits. If you set goals for yourself, however, you can make gradual changes in, and improvements to your writing habits and skills.

To set goals, a first step can be to ask your boss or supervisor for help or support. If your boss has been especially critical of your writing, sit down and *mutually* decide what areas need improvement, how you can pursue improving those skills, and set deadlines for review.

If you don't need or want a boss' input (or if he/she is unwilling to contribute to your goal setting), set goals on your own. Set two or three goals at a time, or set just one goal to improve your writing. Goals can include using greater sentence variety, making outlines for all projects, brainstorming in writing, or cutting out the clutter.

Mark your calendar for two months from now, as a reminder to review whether you have met the goal. During the next two months, make a special effort to achieve that goal every time you write.

At the end of two months, review your goal and ask yourself if you have achieved it, or at least made substantial progress in satisfying the goal. If you haven't improved significantly, then renew your efforts to achieve the goal, and set a new two-month deadline.

When you feel that you have made measurable progress in fulfilling a goal, set a new goal, with another deadline to review yourself in two months. (Remember to mark the calendar each time, so that you don't forget to check your progress.) By continually challenging yourself to improve writing skills, you can continue to improve your professional image and the way others see you.

NOTES

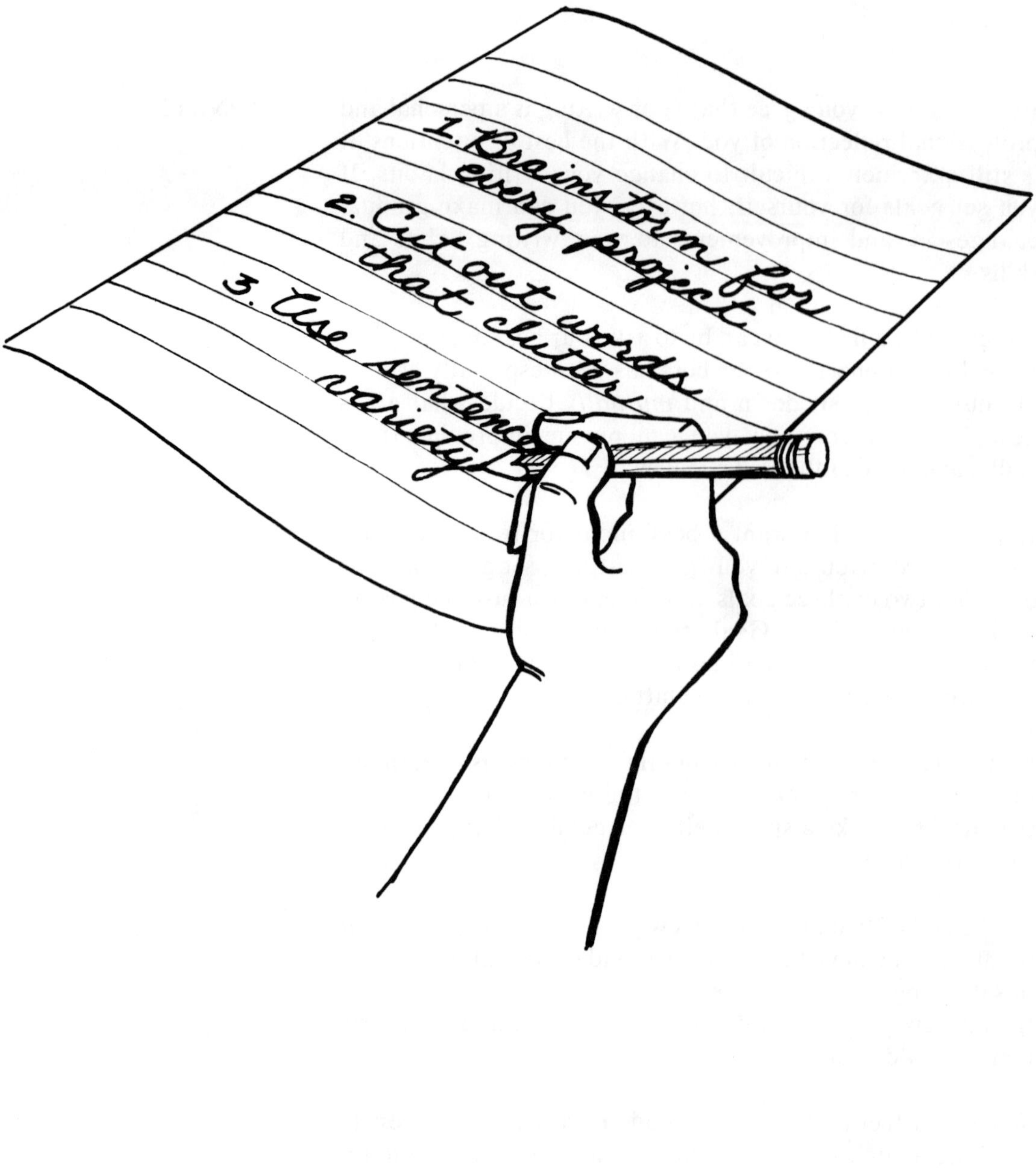

Reading List

Burne, Kevin G., Edward H. Jones, and Robert C. Wylder: *Functional English for Writers,* 1978. Scott, Foresman and Company, Glenview, Illinois.

Graves, Harold F. and Lyne S. S. Hoffman: *Report Writing,* Fourth Edition, 1965. Prentice-Hall, Inc., Englewood Cliffs, New Jersey.

Guth, Hans P.: *Concise English Handbook,* Fourth Edition, 1978. Wadsworth Publishing Company, Belmont, California.

Fowler, H. W.: *A Dictionary of Modern English Usage,* Second Edition, 1983. Oxford University Press, Oxford, England.

Rivers, William E.: *Business Reports,* 1981. Prentice-Hall, Inc., Englewood Cliffs, New Jersey.

Ryckman, W. G.: *The Art of Writing Clearly,* 1983. Dow Jones-Irwin, Homewood, Illinois.

Strunk, William, Jr. and E. B. White: *The Elements of Style,* 1979. MacMillan Publishing Co., New York, New York.

Uris, Auren: *Memos for Managers,* 1975. Thomas Y. Crowell Company, New York, New York.

INDEX

Active voice, 18–19, 24
  in taking minutes, 68
Apostrophe, 75–76
Attachments
  in report writing, 62–64
Audience (reader), 2–3, 7–8
  in letter writing, 24–25, 35, 39–40
  in memo writing, 51, 53
  in machine dictation, 72
  in minute writing, 67
  in report writing, 60–61, 63, 65

Body
  in report writing, 62–64
Books (reference), 5
  reading list, 89–90
Brackets, 76–77
Brainstorm, 7
  as a goal, 87
  in letter writing, 35, 36–37, 38

Clutter (cutting out), 13–17
  setting goal to cut out clutter, 87
Colon, 77
Comma, 77–80
  example of excess, 33

Dash, 80
Deadlines (meeting, setting), 7, 40
Documenting
  background information, 40
  calls in memos, 21
  telephone decisions in memos, 55
Drafts, 7, 9
  editing, 50
  final draft, 35, 36
  for meeting minutes, 69
  in machine dictation, 71

Editing, 9, 12, 24, 47–50
  reports, 65
Ellipsis, 81
Errors (typographical), 24, 47, 65
Exclamation point, 81

Findings
  in report writing, 60–62, 64–65
Formality, 1, 24
  being natural, 7–8
  examples from memo, 51, 55

Goals (setting them), 87–88
Grammar
  example of problem in memo, 56
  punctuation rules, 75–85

Headings and subheadings, 59–60, 64
Hyphen, 81–82

Introduction
  in report writing, 60–61, 63, 65
Italics, 82

Jargon, 3
  list of jargon phrases, 16

Letters
  organizing, 39–45
  planning, 35–38
  pointers for, 21–27
  problem or difficult letters, with
    examples, 27–34

Machine dictation, 71–73
Meeting minutes, 67–69
Memorandums, 1, 51–57

Objectivity, 8, 9
  in problem letters, 39
Organizing
  planning ahead for letters, 38–39
  reports, 63–65
Outline
  as a goal, 87
  for letters, 38
  for machine dictation, 71
  for reports, 64

Paragraphs, 19
Parentheses, 82–83

Passive voice, 18–19, 24
in writing minutes, 68
Period, 83
Personal pronouns, 26–27
use in memo, 57
Planning process, 7
for letter writing, 35–38
Proofreading, 9, 24
also see editing
Punctuation, 75–85
example of extra commas, 33
verbal, in machine dictation, 73
Purpose
in memos, 56–57
in reports, 60
limiting, 36
up front in letters, 23

Question mark, 83
Quotation marks, 83–84
punctuation with, 84

Reader. See AUDIENCE
Recommendations
in report writing, 60–62, 64–65
Reports, 59–65
format for, 59–63
strategy for writing, 63–65
Review. See EDITING
Semicolon, 84–85
Sentences
opening and closing, 25, 43–45
Sentence structure
variety, 11–12, 32–33
variety as goal, 87
variety example in memo, 56
Setting goals, 87–88
"Shun" words, 16–17
Slash, 85
Strategy
for editing, 47–50
for writing letters, 35–41
for writing reports, 63–65
Style (professional), 1–3, 13
professional image, 35, 38
Summary
in report writing, 61, 63, 65
Table of contents, 60, 65
Theme, 36, 38, 39, 42
Thesis. See THEME
Tone, 1, 7–8, 57

Word choice
examples, 33, 42
in problem letter, 39–42
to cut clutter, 9, 13
Writer's block, 5–9, 37